READING IN A
DIGITAL AGE

Charleston Briefings: Trending Topics for Information Professionals is a thought-provoking series of brief books concerning innovation in the sphere of libraries, publishing, and technology in scholarly communication. The briefings, growing out of the vital conversations characteristic of the Charleston Conference and Against the Grain, will offer valuable insights into the trends shaping our professional lives and the institutions in which we work.

The *Charleston Briefings* are written by authorities who provide an effective, readable overview of their topics—not an academic monograph. The intended audience is busy nonspecialist readers who want to be informed concerning important issues in our industry in an accessible and timely manner.

Matthew Ismail, Editor in Chief

READING
IN A
DIGITAL
AGE

DAVID M. DURANT

Published in the United States of America by
ATG LLC (Media)
Manufactured in the United States of America

DOI: http://dx.doi.org/10.3998/mpub.9944117

ISBN 978-1-941269-13-8 (paper)
ISBN 978-1-941269-17-6 (e-book)

against-the-grain.com

One of the effects of living with electric information is that we live habitually in a state of information overload. There's always more than you can cope with.

—MARSHALL MCLUHAN
ON *THE BEST OF IDEAS* ON CBC RADIO IN 1967

CONTENTS

ACKNOWLEDGMENTS

I would like to thank Katina Strauch and everyone involved with the *Charleston Briefings* series for giving me this opportunity. In particular, I wish to thank my editor, Matthew Ismail, for his very helpful guidance as well as his extreme patience as I struggled to complete this manuscript. Finally, I wish to acknowledge all my colleagues with whom I have had the opportunity to discuss this topic and hone my ideas. I have had a number of fruitful and interesting conversations in locations ranging from the Charleston Conference to my own institution, all of which enabled me to develop, deepen, and clarify my thoughts on the future of reading in this digital age. I especially would like to thank Tony Horava of the University of Ottawa, with whom I have previously collaborated on this vitally important topic. All of these people have helped me tremendously and deserve credit for whatever positive contribution this volume makes to the discussion. Responsibility for all faults are, of course, mine alone.

INTRODUCTION

How we read has dramatically changed with the advent of the digital age. Just twenty years ago, we still primarily read items in print format: books, magazines, newspapers, and academic journals. Now in the web environment, our reading options have enormously expanded: web pages, blogs, tweets, Facebook posts, texts, e-mails, and e-books. In the words of Canadian librarian Barry W. Cull, "The Internet is a text-saturated world. It could only have succeeded in a highly literate society."[1]

At first glance, this shift would suggest that reading is thriving like never before in the twenty-first century. After all, more texts—from tweets to full monographs—are now accessible to more people than ever before. Yet in recent years, numerous authors, both popular and scholarly, have written books and articles expressing deep concern about the impact of e-reading on our ability to read, write, and even think. I myself have expressed these concerns in my own writings.[2] Others, who champion the transition to digital reading, believe that any costs in terms of literacy will be minimal or may even prove to be net gains.

So why has the future of reading, and digital reading in particular, become such a tremendous source of controversy? This briefing will explore the arguments regarding the nature and consequences of reading in the digital environment. It will also examine some of the key scientific studies analyzing the potential differences between print and digital reading as well as comparative sales trends and surveys of reader behavior and preferences. Are we truly headed for a mostly digital reading environment? Will print make a resurgence? Or will there be a hybrid multiformat reading future? Finally, I will consider the key question: How do we, as librarians, publishers, and software creators, work to preserve reading in all its richness in the digital age?

Before answering these questions, though, the first order of business is to step back and think about reading in general.

One point worth noting as I begin is that this briefing focuses primarily on North America and Western Europe. This is where digital reading has made the greatest strides and where the debate over the future of reading is being waged most earnestly. Thus I focus my attentions here for purposes of clarity and brevity. The state of reading in other parts of the world will inevitably be shaped by the unique factors of each region. While I believe this briefing will be relevant to readers from those areas as well, these distinctions are worth keeping in mind.

READING AND NEUROPLASTICITY

There are several key points to keep in mind when we talk about reading. One is that the ability to read is not innate—that is, we are not born able to read. It is a learned skill. The human brain is not designed for reading; rather, reading developed as a result of a phenomenon called neuroplasticity. In the words of Maryanne Wolf and Mirit Barzillai, "Plasticity enables the brain to form new connections among the structures underlying vision, hearing, cognition, and language."[3] Reading, in effect, was made possible by the brain's ability to rewire itself. The more one reads, the more deeply the neural pathways that facilitate reading take hold. The opposite is equally true. I'll be coming back to this point later.

LINEAR VERSUS
TABULAR READING

It is also important to note that all reading is not the same. Reading Tolstoy is not the same as reading a tweet or a restaurant menu. Communication scholar Christopher Rowe has divided reading into two types, linear and tabular: "Linear or intensive reading characterizes the way we consume narrative fiction. The tabular mode of reading is interrogative, seeking information about a specific subject."[4] Along the same lines, N. Katherine Hayles, a professor at Duke University, has discerned three basic forms of reading: close, hyper, and machine. Hayles's concept of close reading corresponds to what Rowe defines as linear reading, while hyper reading is roughly analogous to what he calls tabular reading. As Hayles notes, both forms of reading are valuable and, in fact, complement each other: "Close and hyper reading operate synergistically when hyper reading is used to identify passages or to home in on a few texts of interest, whereupon close reading takes over. . . . Skimming and scanning here alternate with in-depth reading and interpretation."[5]

Hayles's third type of reading, machine reading, consists of "human-assisted computer reading, that is, computer algorithms used to analyze patterns in large textual corpora where size makes human reading of the entirety impossible."[6] In other words, machine reading is everything from using the "Find" command in a PDF document to using textual analysis software.

Leaving aside machine reading, whether we say linear and tabular or close and hyper (we will use "linear" and "tabular" for the remainder of this briefing), it is clear that there are two main forms of reading that each foster very different types of intellectual abilities. Linear reading involves the ability to read an extended narrative in continuous, in-depth fashion and reflect upon its meaning. Wolf and Barzillai define linear reading as "the array of sophisticated processes that propel comprehension and that include inferential and deductive reasoning, analogical skills, critical analysis, reflection, and insight. The expert reader needs milliseconds to execute these processes; the young brain needs years to develop them."[7] This form of reading depends on and in turn helps foster skills such as sustained focus and attention, deep concentration, and the ability to memorize information and integrate it into conceptualized forms of knowledge and self-awareness.[8]

In contrast, tabular reading focuses on either reading short pieces of text or browsing or skimming texts in search of specific pieces of information. Examples include browsing a web page or looking up a word in a dictionary. Tabular reading thus tends to be nonlinear in nature, develops rapid pattern recognition and quick decision making, and is often interactive instead of solitary.[9]

This distinction between linear and tabular reading has come to be central to the discussion over reading in the digital age.

PRINT READING

To a certain extent, the debate over the future of reading is simply a continuation of previous arguments concerning the impact of new technologies on society. As far back as 1934, scholar Lewis Mumford expressed worries about the spread of technology in his *Technics and Civilization*. In the 1960s, Canadian media theorist Marshall McLuhan famously observed that "the medium is the message." More specifically, he predicted in his 1962 work *The Gutenberg Galaxy* that the growth of visual media such as films and television would greatly affect our ability to absorb and communicate via the written word. In the 1970s and 1980s, New York University (NYU) communication professor Neil Postman wrote widely on the negative impact of screen-based technologies, primarily television, which he believed were reducing people's attention spans and ability to think.

The first major work to express concern about how digital text would alter the nature of reading was Sven Birkerts's 1994 *The Gutenberg Elegies*. Published at the dawn of the World Wide Web and influenced by previous technology critics such as McLuhan, Birkerts warned of the dramatic impact that hypertext might have on the reading experience: "Words read from a screen or written onto a screen—words which appear and disappear, even if they can be retrieved and fixed into place with a keystroke—have a different status and affect us differently from words held immobile on the accessible space of a page. But McLuhan's analysis of the print-to-electronics transformation centered upon television and the displacement of the printed word by transmissions of image and voice. But what about the difference between print on a page and print on a screen? Are we dealing with a change of degree, or a change of kind?"[10]

Birkerts's question—Does digital text differ merely in degree from print reading, or does it represent a far more transformative change?—lies at the heart of the current controversy over the future of reading. As the web exploded in popularity in the late 1990s and as "Web 2.0" emerged in the early twenty-first century, the general assumption was that the impact of digital text was mostly, if not entirely, positive. Despite the occasional effort to raise the alarm, concerns about the effect that the spread of screen-based reading was having on us, both individually and as a society, were fairly muted. In 2008, however, an attention-grabbing article would succeed in bringing these concerns front and center, and the debate over the future of reading would be joined in earnest.

Nicholas Carr, a technology writer and anything but a Luddite, published an article in *The Atlantic* titled "Is Google Making Us Stupid?" The piece served as the digital age's metaphorical equivalent of the nailing of the Ninety-Five Theses to the church door at Wittenberg. Not only did Carr's essay inaugurate in earnest the twenty-first-century reading debate; it is in many ways the urtext of digital reading skepticism, laying out the essential arguments that most critics of online reading have since employed. As such, the essay is worth quoting at length.

Carr begins his piece by describing how, following the advent of digital reading, he has much greater difficulty in reading at length and in depth:

> I'm not thinking the way I used to think. I can feel it most strongly when I'm reading. Immersing myself in a book or a lengthy article used to be easy. My mind would get caught up in the narrative or the turns of the argument, and I'd spend hours strolling through long stretches of prose. That's rarely the case anymore. Now my concentration often starts to drift after two or three pages. I get fidgety, lose the thread, and begin looking for something else to do. I feel as if I'm always dragging my wayward brain back to the text. The deep reading that used to come naturally has become a struggle.[11]

In looking for an explanation, Carr noted, "For more than a decade now, I've been spending a lot of time online, searching and surfing and sometimes adding to the great databases of the Internet."[12] While acknowledging that the web had been of immense benefit to him as a writer, he expressed the fear that the advantages of online text have come with a price tag. As he put it, "What the Net seems to be doing is chipping away my capacity for concentration and contemplation. My mind now expects to take in information the way the Net distributes it: in a swiftly moving stream of particles. Once I was a scuba diver in the sea of words. Now I zip along the surface like a guy on a Jet Ski."[13]

Carr's explanation for these cognitive changes, both in himself and in others he had spoken to who had experienced the same phenomenon, was neuroplasticity (though the word itself does not appear in the article): "The human brain is almost infinitely malleable. People used to think that our mental mesh-work, the dense connections formed among the 100 billion or so neurons inside our skulls, was largely fixed by the time we reached adulthood. But brain researchers have discovered that that's not the case."[14]

Carr would go on to expand on this argument in his 2010 book *The Shallows*, noting that there is a growing body of research-based and anecdotal evidence that reading from a printed page is different than reading from an electronic screen.[15] In this view, print books and e-books facilitate two very different types of reading. Whereas deep print reading tends to foster sustained attention and in-depth reflection, e-reading fosters impatience and a need for immediate gratification. E-reading is also much more likely to be prone to distraction, as it is often done on devices that also offer e-mail, apps, or access to the Internet, which in Carr's words, "seizes our attention only to scatter it."[16] Thus screen-based reading is often much less conducive to memorization than print reading.

Carr's argument is supported by numerous studies—ranging from scientific eye-tracking research to usage analysis to surveys of readers—showing that people reading in digital format are far more likely to engage in a form of superficial power browsing or skimming than they are to read in depth. For example, in 2009, a team of researchers at the University of California, Los Angeles (UCLA) found that Internet searching activated many more areas of the brain than did reading text from a page.[17] While at first this sounds like a point in favor of e-reading, this is not necessarily the case. Instead, this increased brain activity likely reflects the stimulative, distraction-laden nature of screen reading that actually impairs the ability to memorize, reflect, and absorb in the way that print texts, conducive to intensive linear reading, allow.[18] Web usability pioneer Jakob Nielsen has likewise found that users do not read web pages in a linear manner but rather scan them using what he has called an "F-shaped pattern," making shorter and less intensive scans of text the farther the user goes down the page.[19] A 2008 British Library analysis found that "It is clear that users are not reading online in the traditional sense. It almost seems that they go online to avoid reading in the traditional sense."[20]

For digital skeptics, the key to understanding this transformation in how people are reading lies in the concept of neuroplasticity. The more we read from screens in tabular fashion, the more our brains rewire themselves to facilitate this activity and the harder it becomes to engage in deep print reading. Format does matter. Text is not interchangeable. While e-reading certainly has its advantages, it is not the same as reading from the printed page. It fosters a different set of cognitive skills and a qualitatively different way of thinking. Digital skeptics argue, in short, that the rise of e-reading has fostered tabular reading at the expense of linear reading, and thus it has greatly increased our ability to access information at the expense of our ability to convert it into conceptual knowledge.

THE MORE WE READ FROM SCREENS IN TABULAR FASHION, THE MORE OUR BRAINS REWIRE THEMSELVES TO FACILITATE THIS ACTIVITY AND THE HARDER IT BECOMES TO ENGAGE IN DEEP PRINT READING.

While Carr is the most widely known skeptic of digital reading, many other authors, both popular and scholarly, have expressed similar concerns about the impact of screen-based reading. Wolf and Barzillai have argued that "the digital culture's reinforcement of rapid attentional shifts and multiple sources of distraction can short-circuit the development of the slower, more cognitively demanding comprehension processes that go into the formation of deep reading and deep thinking. If such a truncated development occurs, we may be spawning a culture so inured to sound bites and thought bites that it fosters neither critical analysis nor contemplative processes in its members."[21]

Hayles has reported encountering many of the same concerns, not just in her research, but in her broader work in academia:

Anecdotal evidence hooked me on this topic five years ago. Everywhere I went, I heard teachers reporting similar stories: "I can't get my students to read long novels anymore, so I've taken to assigning short stories"; "My students won't read long books, so now I assign chapters and excerpts." I hypothesized then that a shift in cognitive modes is taking place, from the deep attention characteristic of humanistic inquiry to the hyper-attention characteristic of someone scanning Web pages. Since then, the trend has become even more apparent, and the flood of surveys, books, and articles on the topic of distraction is now so pervasive as to be, well, distracting.[22]

Similarly, Ferris Jabr, writing in 2013 for *Scientific American*, offered this sympathetic summary of the digital skeptic's case against screen-based reading: "Even so, evidence from laboratory experiments, polls and consumer reports indicates that modern screens and e-readers fail to adequately recreate certain tactile experiences of reading on paper that many people miss and, more importantly, prevent people from navigating long texts in an intuitive and satisfying way. In turn, such navigational difficulties may subtly inhibit reading comprehension. Compared with paper, screens may also drain more of our mental resources while we are reading and make it a little harder to remember what we read when we are done."[23]

More recently, Naomi Baron, a scholar of linguistics at American University, has expressed many of these same concerns regarding the impact of screen-based reading. In her 2015 book *Words Onscreen: The Fate of Reading in a Digital World*, Baron has argued that "one of the major effects of digital screens is to shift the balance from continuous reading to reading on the prowl."[24] As part of the research for her book, in 2013 Baron surveyed a select sample of undergraduates in the United States, Germany, and Japan regarding their reading habits and preferences. When she asked them about multitasking, 85 percent of the American students reported multitasking while reading on a screen versus 26 percent who multitasked while reading in hard copy. Results among the German students were comparable.[25]

Concerns about the impact of digital text on our ability to read in depth have gained such resonance that the topic has been picked up by major media outlets. An April 2014 *Washington Post* piece, for example, worried that "humans seem to be developing digital brains with new circuits for skimming through the torrent of information online. This alternative way of reading is competing with traditional deep reading circuitry developed over several millennia."[26] Among the experts quoted in the article is Wolf, who told the *Post* that "I worry that the superficial way we read during the day is affecting us when we have to read with more in-depth processing."[27]

Digital reading skeptics are especially worried not just about the present state of reading but about what might happen to linear reading in the future. British neuroscientist Susan Greenfield, in her 2015 book *Mind Change*, expressed her concern that "these powerful interactive screen technologies are not just exciting experiences but critical tools that have reshaped our cognitive processes and will continue to do so, creating both benefits and problems. The difference between silicon and paper, the distractions of multitasking and hypertext, and the tendency to browse rather than to think deeply all suggest fundamental shifts in how our brains are now being asked to work."[28]

The ultimate worry is about what will happen if today's children are only exposed to screen-based reading. Given what we know about neuroplasticity, this change could result in their failure to develop the ability to engage in immersive linear reading. Wolf summarized this concern in a 2010 article: "The reading circuit's very plasticity is also its Achilles' heel. It can be fully fashioned over time and fully implemented when we read, or it can be short-circuited—either early on in its formation period or later, after its formation, in the execution of only part of its potentially available cognitive resources."[29]

DIGITAL READING

Like Carr, most digital skeptics freely acknowledge the benefits of screen-based reading. It has made more text readily available than ever before in human history. It has made reading a far more mobile and portable activity. Instead of lugging around a handful of books, you can have thousands of e-books at your fingertips on a Kindle, tablet, or smartphone and can access them virtually anywhere. Thanks to social media, reading can now be much more of a shared, interactive process and potentially that much more interesting as a result. Finally, the digital reading environment has, in many cases, made tabular reading a much easier exercise than it was before. Features such as the "Find" command make it much simpler to find specific pieces of information in larger texts, while online search engines have obviated much of the need for print dictionaries and encyclopedias.

Again, most digital skeptics both accept these points and regard these phenomena as substantially positive. They are not opposed to digital texts in principle and are certainly not against tabular reading. Indeed, they regard both as indispensable. It is not even reading on digital devices that they necessarily oppose. The primary fear of those concerned about the spread of digital reading is the possibility that by transitioning to an almost exclusive reliance on reading from digital devices without thinking through the matter, we risk losing much if not all of our ability to read complex, linear texts at length. Through such means as neuroplasticity and the great potential for distraction built into many digital devices, the online digital environment could well be fostering tabular reading while eroding our ability to engage in deep linear reading. If this is the case, the way in which we read, write, and even think is changing enormously.

At the risk of oversimplifying things, those who reject the concerns offered by the digital skeptics fall into two main camps. The first school of thought rejects the notion that transitioning to reading primarily in digital format will have a major impact on how people read. The second school of thought actually agrees with the skeptics that digital text will dramatically transform reading, but they argue that this will prove to be a good thing.

CONTINUITY OF DIGITAL WITH PRINT?

Many among the first school of digital defenders make the case, contra McLuhan, that format is essentially irrelevant: text is interchangeable whether it appears on a printed page, a computer screen, or a Kindle. In a September 2010 piece for the *Chronicle of Higher Education*, Jeffrey R. Di Leo, a dean at the University of Houston at Victoria, argues that "academe must transform itself from a fundamentally print culture to one that is fundamentally digital" and openly looks forward to the day when "the myth of the book will be overcome."[30] As Di Leo puts it, "There is nothing intrinsically inferior about spreading knowledge on a screen rather than on a printed page, and plagiarism is an ethical issue, not a material one. Words may look better in print, and a book may feel better in your hands than a Kindle or an iPad, but the words are the same."[31]

Writing in the same publication, publishing executive Diane Wachtell argues, "We do not need books."[32] In her view, long-form texts are what matter, and the precise container is unimportant: "We are mistaking the package for the thing itself. What is crucial at a time when habits of consumption are changing—for reasons both economic and technological—is to ensure the future of lofty ideas, whether they are set in Bodoni or pixels, hand-sewn at the binding or backlit and scrolled."[33]

Among the leading critics of the case against digital reading is *New York Times* technology writer Nick Bilton. In his 2010 book *I Live in the Future & Here's How It Works*, Bilton attributes much of the worry over the impact of screen-based reading to a phenomenon he calls "technochondria": "Fear of the new and fear of the unknown."[34]

Bilton is unconvinced by concerns over the possible rewiring of our brains induced by digital reading. If anything, Bilton argues, neuroplasticity will work in our favor: "Just as well-meaning scientists and consumers feared that trains and comic books and television would rot our brains and spoil our minds, I believe many of the skeptics and worrywarts today are missing the bigger picture, the greater value that access to new and faster information is bringing us. For the most part, our brains will adapt in a constructive way to this online world."[35] Why does Bilton believe this? "Because we've learned how to do so many things already, including learning how to read."[36]

Similarly, Clive Thompson of *Wired Magazine* argues that the belief that print reading fosters superior attentiveness compared to digital reading is primarily a result of deeply held cultural prejudices that create a self-fulfilling prophecy. He explained his views in a 2015 piece discussing his ultimately successful attempt to read *War and Peace* on his smartphone: "But what happens if we treat digital screens with the same romance, the same intensity of focus? Studies suggest that the cognitive distinctions go away: We learn just as much, and retain just as much, as we do on paper. When we believe that reading on a phone is equally 'serious' as reading on paper, we internalize that reading just as deeply."[37]

Other defenders of digital reading have echoed this argument. In a July 2015 review of Baron's *Words Onscreen*, John Jones, a professor of writing at West Virginia University, makes the case that the perceived inferiorities of screen-based reading are due to a mix of the cultural predilections emphasized by Thompson along with the limitations of current digital reading devices. In his view, both aspects will sort themselves out as digital reading continues to develop: "Rather than arguing for a return to print for serious reading or demonstrating that 'digital reading' is inherently flawed, what anecdotes of our difficulties adjusting to the various forms of reading on our screens suggests is that we are still at an early stage in the development of digital reading tools. . . . More importantly, we are not yet cultured to digital reading as we are with reading print—we are still training ourselves to manage the new distractions produced by our devices and becoming literate in the navigational affordances of digital texts."[38]

One subset of the argument that digital reading is not inherently different from print reading relies on dedicated e-reading devices, such as the Kindle and the Nook. Unlike most digital devices, e-readers are designed to mimic the experience of print reading as closely as possible. In the opinion of some, dedicated e-readers offer the best of both worlds: a digital reading technology that preserves the key features of deep print reading. Alan Jacobs, a professor of English at Baylor University, has written how he too—like Carr—found himself losing the ability to read lengthy linear narratives. However, Jacobs regained the ability to engage in deep linear reading once he purchased a Kindle. He soon found his "ability to concentrate . . . restored almost instantly."[39] In Jacobs's view, "E-readers are by any measure far less distracting than an iPad or a laptop. It's at least possible for new technologies to be part of the solution instead of part of the problem."[40]

If textual content is all that matters and one format is as good as another, then it only makes sense that reading should become a primarily digital affair for all the reasons of space, portability, and ease of access discussed above. If format is truly irrelevant, then print can safely be relegated to niche status or even abandoned altogether. It is not a question of if, but when. This belief, implicit in the arguments of many supporters of digital reading, has been made explicitly by some. Purdue University librarian George Stachokas, for example, has argued not only that mainly electronic libraries are inevitable but that "this transition could be completed in five to ten years in most academic libraries in North America, the UK, Australia, and New Zealand."[41] Current print-electronic hybrid library collections, in this view, are merely a short-term product of circumstances that will soon be overcome. In a September 2015 analysis of recent sales trends in publishing, Matthew Ingram states that "digital sales are going to increase, and print is likely to become a niche market over time."[42] In early 2016, digital publishing consultant Mike Shatzkin told the BBC that the death of print is "inevitable."[43] In his words, "I think there will come a point where print just doesn't make a lot of sense."[44]

Technology author Marc Prensky has even called for college campuses to go completely "bookless," in the "sense of allowing no physical books."[45] In his vision, students caught in possession of print books would have them confiscated and replaced with access to an electronic version of the same title. As Christine Rosen noted back in 2008, "Digital literacy's advocates increasingly speak of replacing, rather than supplementing, print literacy."[46]

DIGITAL BETTER THAN PRINT?

The second set of proponents of digital reading actually agrees with the likes of Carr, Baron, and Wolf that e-reading is substantially different from reading in print. Where they differ is that they see this as a generally positive development. For example, Clay Shirky, a communication scholar at NYU and champion of new media, has expressed the view that we have entered a new age of "information abundance," in which the digital environment will enable more people to produce more content than ever before. In Shirky's view, the print codex and the type of reading and thinking it fosters are merely byproducts of the technology of the printing press and will rightly be superseded by new cultural forms produced by digital media.[47] In a 2013 online exchange with Carr, Shirky predicted that "the experience of reading books will be displaced by other experiences" and pronounced himself "quite cheerful about the ongoing destruction of pre-digital patterns of life, because I think something better will come from it."[48]

Wired Magazine founder Kevin Kelly likewise believes that the digital information environment will create something superior to the stable, tangible print codex. In a 2010 essay on the differences between print and screen reading, Kelly essentially flipped the argument of digital skeptics on its head, embracing the changes brought about by digital text as a form of progress: "Books were good at developing a contemplative mind. Screens encourage more utilitarian thinking. A new idea or unfamiliar fact will provoke a reflex to do something: to research the term, to query your screen 'friends' for their opinions, to find alternative views, to create a bookmark, to interact with or tweet the thing rather than simply contemplate it. Book reading strengthened our analytical skills. Screen reading encourages rapid pattern-making, associating this idea with another, equipping us to deal with the thousands of new thoughts expressed every day. The screen rewards, and nurtures, thinking in real time."[49]

More recently, Robert Stein, founder of the Institute for the Study of the Book, believes that the shared elements of the digital reading environment make it superior to the solitary nature of print reading: "Why would you want to read by yourself if you can have access to the ideas of others you know and trust, or to the insights of people from all over the world?"[50]

In the view of many such unabashed supporters of digital reading, the book as a discrete linear entity will likely disappear. Kelly, in a famous 2006 essay, noted that "once digitized, books can be unraveled into single pages or be reduced further, into snippets of a page."[51]

As you would expect, digital skeptics are much less sanguine at this prospect. Carr, for example, has emphasized the importance of the book as a discrete physical entity in contrast to the amorphous, indistinct nature of electronic information on the Internet. In his view, "An electronic book is therefore a contradiction in terms. To move the words of a book onto the screen of a networked computer is to engineer a collision between two contradictory technological, and aesthetic, forces."[52] Academic librarian Jeff Staiger has echoed these concerns, warning that "it may be that by dematerializing the book and making its wholeness invisible and intangible, the e-book weakens the very boundaries and concept of the book, making it that much easier to think of the book as a mere fount of textual bits."[53]

WHAT THE RESEARCH TELLS US

How, then, is one to unravel the conflicting claims regarding the benefits of reading in print format versus digital format? One obvious starting place is by looking at the various studies that have been done comparing reading the printed page to reading on digital devices. We've already mentioned a few earlier, but the number of such projects has grown tremendously as this topic has become a source of controversy.

On the one hand, a number of research studies have appeared that describe little or no difference between reading on screens and reading on paper. For example, in 2010, usability expert Jakob Nielsen did a study comparing reading speed and comprehension for a printed book, iPad, Kindle, and PC. The study found that users read somewhat faster in print, while comprehension was similar regardless of platform. User satisfaction levels were comparable for the print book, Kindle, and iPad.[54]

A 2011 study by Johannes Gutenberg University in Germany used eye tracking and EEG readings of more than fifty subjects, a mix of young adults and senior citizens, to compare the mental effort required to read from a print book, an e-reader, and a tablet. What they found was that there was no real difference between the three formats in terms of comprehension or effort expended. As the researchers put it in the 2013 article describing their study, "The present findings provide no evidence to support the assumption that online reading effort increases when people read on digital devices as opposed to paper."[55] One really interesting part about the Gutenberg University study is that, according to the researchers, "both younger and older Participants showed an overwhelming preference for the book page when asked to choose their preferred reading medium." However, "though participants stated that they preferred the book page over the electronic reading devices, none of the quantitative online measures collected support that reading was more effortful for the digital media."[56] The authors offer the following explanation, one very much in keeping with the preexisting cultural prejudices theory favored by Thompson and Jones: "This suggests that the overwhelming public opinion that digital reading media, though convenient, reduce the pleasure of reading is a cultural rather than a cognitive phenomenon. From this perspective, the subjective ratings of our participants (and those in previous studies) may be viewed as attitudes within a period of cultural change."[57]

There have been several other studies that have also found no real difference in comprehension between reading from a print codex and a digital device. Most notably, a study published in 2013 by Sara Margolin, a professor of psychology at the College at Brockport, State University of New York, analyzed reading comprehension among ninety undergraduates, of whom a third read ten short passages on paper, a third the same passages on a computer, and the final third on a Kindle. According to Margolin and her coauthors, "The results indicated no significant differences among media presentation types. This lack of significant differences in comprehension accuracy across media platforms indicates that if comprehension differences exist, the present research did not find them and therefore are likely to be very small differences or at least moderated by some other factor."[58]

While some studies reinforce the idea that there is no inherent difference in print versus screen-based reading, and thus whatever differences exist are the result of ingrained cultural practices, others support the arguments of digital skeptics. Most prominent in this regard has been the work of Norwegian researcher Anne Mangen. Mangen, who works at Stavanger University in her native Norway, has done extensive work comparing print versus digital reading. In one study published in 2013, Mangen and two colleagues worked with a group of seventy-two Norwegian tenth graders. Half of these students read two texts in print format and the other half read the same two texts in PDF format on a computer. Both groups were tested afterward on comprehension. Mangen and her colleagues found that "subjects who read the texts on paper performed significantly better than subjects who read the texts on the computer screen."[59]

Mangen and her colleagues have followed up this study with another as-yet-unpublished one. This latter project used a similar methodology to the previous one. A group of fifty readers were asked to read a twenty-eight-page story by Elizabeth George, half in paperback and half on a Kindle. The results, released by Mangen at a July 2014 conference in Italy, revealed that the Kindle readers did significantly worse when tested about the plot of the story. The British newspaper *The Guardian* quoted Mangen discussing her findings:

> When you read on paper you can sense with your fingers a pile of pages on the left growing, and shrinking on the right. . . . You have the tactile sense of progress, in addition to the visual. [The differences for Kindle readers] might have something to do with the fact that the fixity of a text on paper, and this very gradual unfolding of paper as you progress through a story, is some kind of sensory offload, supporting the visual sense of progress when you're reading. Perhaps this somehow aids the reader, providing more fixity and solidity to the reader's sense of unfolding and progress of the text, and hence the story.[60]

According to the article, Mangen's research has left her very much concerned about the impact of screen-based reading on our ability to engage in deep linear reading.[61] So a look at the available research studies on print versus digital reading reveals that some refute the arguments of the digital skeptics, while others support them. After all this, we are right back where we started. One of these sets of studies must be wrong; they can't both be right, can they? Surely someone has raised legitimate questions regarding at least some of these studies?

There have, in fact, been commentators who have cautioned us against drawing sweeping conclusions from what are relatively limited studies. Hayles, in discussing Carr's *The Shallows*, argues that "although Carr's book is replete with many different kinds of studies, we should be cautious about taking his conclusions at face value."[62] As Jeremy Greenfield of *Forbes* pointed out, only two of the students in Mangen's Elizabeth George study had used Kindles before, while all of her students were undoubtedly quite familiar with print, thus possibly skewing the results. (In all fairness, Mangen and her coauthors noted this themselves in their article.)

Greenfield noted that this might indicate a broader problem with studies that suggest major differences in how people read digital text as opposed to how they read from paper: "Experiments have been conducted for decades comparing reading on screens versus reading in print. Early results suggested very strongly that print had massive advantages. But as time went on, screen reading drew closer to print reading. One theory is that test subjects were more used to the medium as more people read on screens, and that this changed the findings. E-reading is such a new technology, and most of the research subjects part of the digital group were unfamiliar with it and with the Kindle device, as opposed to reading in print, which nobody is unfamiliar with."[63]

So perhaps Bilton, Thompson, and Jones are correct that there is no fundamental difference between print and screen-based reading. Perhaps the studies to the contrary simply are the result of user unfamiliarity with the technology, cultural prejudices in favor of print, and flaws with current e-reading technology that will eventually be corrected. Unfortunately, there are also limitations to the studies that are supportive of digital reading that should be pointed out as well.

For example, both the Gutenberg University and Margolin studies involved short pieces of text. The readings in the latter were about five hundred words; in the German study, fewer than three hundred. One must wonder what the results would have been had the subjects been required to read five thousand words, let alone fifty thousand. Would the comprehension scores still have been similar? In addition, one of the main concerns regarding digital reading is that much of it is done on multipurpose devices, such as tablets and smartphones, which offer numerous diversions more immediately stimulating than immersing yourself in a text. Again, how would the subjects in studies such as Margolin's and Gutenberg University's have fared had they not been reading in a controlled environment? Would they have been as diligent and able to avoid distraction if they hadn't been essentially a captive audience with graduate assistants looking over their shoulders or electrodes attached to their heads?

So if the scientific research on each side of the debate has its limits, where are we to turn? Fortunately, there are other forms of empirical evidence that we can use to get a sense of this issue. In this case, we can rely on what readers themselves have to say about whether they prefer print or digital and look at publishing sales and usage figures as well as user preference surveys.

WHAT DO READERS SAY?

The growth of digital reading has been rapid and undeniable. According to a January 2014 Pew Internet study, 28 percent of all Americans sixteen or older had read an e-book in the previous twelve months, up from a figure of 16 percent in late 2011 and 23 percent at the end of 2012. Forty-two percent of Americans owned a tablet, up from 10 percent in late 2011 and 24 percent in late 2012, while 32 percent owned a dedicated e-reading device such as a Kindle or Nook—a major increase from the 19 percent figure for 2012. In all, 50 percent of Americans owned either a tablet or a designated e-reader.[64]

The same Pew study offered insights as to which e-reading devices were being used for e-books. Among all those who read at least one e-book, 57 percent used a dedicated e-reader, 55 percent used a tablet, and, interestingly, 32 percent used a smartphone.[65]

E-book sales figures further illustrate this trend. In 2011, Amazon announced that its e-book sales had exceeded its print sales.[66] In 2012, a survey of American publishers revealed that e-books made up 20 percent of the trade market, with 457 million e-books sold during the year, a massive increase on the ten million sold in 2008.[67]

Library circulation figures add yet more clarity to the overall picture. In a 2011 piece for *Library Journal*, Rick Anderson of the University of Utah analyzed circulation rates per student at ten Association of Research Libraries member institutions. In his view, his findings indicate that "the trend away from print books is even more pronounced than we've often understood or assumed."[68]

THE PLATEAUING
OF E-BOOK SALES

Based on these trends, many have assumed that we are well on our way to
an all-digital future. A deeper look at the data, however, indicates that such
predictions might well be premature. For example, recent sales data show
that the growth of e-books has substantially slowed in the last several years.
According to an August 2013 study by the Book Industry Study Group, sales
of new e-books have leveled off at 30 percent of overall book unit sales and
about 15 percent of dollar sales. The same study showed that the percentage of book buyers who have bought an e-book has stagnated at around
25 percent. More recently, the *New York Times* reported on September 22,
2015, that e-book sales declined 10 percent in the first five months of 2015,
according to the Association of American Publishers (AAP), while "digital
books accounted last year for around 20 percent of the market, roughly the
same as they did a few years ago."[69] While sales of independently published
e-books may have balanced out declining sales from major publishers, the
Times also noted that many independent booksellers are seeing increased
print sales, driven in part by a return to print among many of their cus-
tomers. And while hardcover sales may be lagging, paperback sales grew by
8.4 percent in the first five months of 2015.[70] As Jeremy Greenfield noted
at Digital Book World, "E-books have stalled out on their way up to higher
altitude."[71]

Sales figures for 2016 have further confirmed this trend. AAP figures for
January–July 2016 show that sales of paperback books grew 8.4 percent over

the same period in 2015, and hardback sales increased by 2.6 percent. E-book sales, in contrast, declined by 19.2 percent.[72]

Some observers have questioned the significance of this trend. Ingram, in a piece for *Fortune*, argues that the trend of falling e-book sales is deceptive, as it does not reflect independently published non-AAP e-books, whose sales have increased in large part because they tend to be much cheaper than e-books published by mainstream publishers.[73] His point is reinforced by news that Amazon reported an increase in its e-book sales in 2015.[74] In a similar vein, Andrew Richard Albanese has argued in *Publishers Weekly* that "just as the e-book market started to boom, the major publishers put a collective thumb on the scales to tip readers back toward print, with efforts that included a scheme with Apple to raise e-book prices, and burdensome restrictions on library e-books."[75] However, even factoring in an increase in major-publisher e-book prices, if users truly preferred e-books as a rule, this should not make that big a difference. After all, the other advantages of e-books (portability, immediacy of purchase, etc.) still apply. In other words, if major-publisher e-book sales are this price sensitive, this would indicate that many users do indeed prefer print if prices are comparable. As the *Times* noted in September 2015, "With little difference in price between a $13 e-book and a paperback, some consumers may be opting for the print version."[76] The dramatic growth in the popularity of e-books, followed by their equally dramatic leveling off, strongly suggests that the e-book market has found its level for now.

This is not just the case in the United States. As the BBC reported on August 14, 2015, British e-book sales have likewise plateaued, leveling off at a 30 percent market share. Overall, British e-book sales amounted to 393 million pounds in 2014 versus 1.7 billion pounds worth of print sales. In fact, two major U.K. booksellers actually saw a year-over-year increase in their print book sales during the 2014 holiday season.[77] Continuing this trend, the five largest U.K. publishers saw a decline in e-book sales in 2015, with a cumulative average drop of 2.4 percent. While this trend is somewhat offset by increased sales of independent e-books, as in the United States, this does not seem to be enough to disprove the overall phenomenon. As Philip Jones, editor of the U.K. magazine *The Bookseller*, told *The Guardian*, "I think overall, the digital market has certainly gone up, if you include smaller

"E-BOOKS HAVE STALLED OUT ON THEIR WAY UP TO HIGHER ALTITUDE."

publishers and self-published books and digital-only publishers. But I don't think that changes the overall picture of the ebook marketplace, which has slowed down from 2012 to 2014, and which will, I think, continue to slow as readers migrate from dedicated e-ink reading devices to tablets and mobile phones."[78]

At the same time that e-book sales have stalled, sales of dedicated e-readers such as Kindles and Nooks have actually declined. According to the BBC, "Kindle sales—peaking at 13.44 million in 2011—fell back to 9.7 million in 2012 and have plateaued since. Barnes & Noble's Nook e-reader has been losing about $70m (£45m) a year and the US bookseller has been trying—and failing—to find a buyer for the division."[79]

To be honest, the decline in e-reader sales is likely due not only to reduced e-book sales but also to a change in how people are reading e-books. The natural tendency in the digital era has been for single-purpose devices to be relegated to boutique status by general multipurpose devices, such as digital cameras being superseded by smartphones. This now appears to be happening in the digital reading environment. Global e-reader sales are estimated to have fallen 36 percent from 2011 to 2012, going from twenty-three million units sold to fewer than fifteen million. The September 2015 *Times* piece, citing data from Forrester Research, reports that e-reader sales have dropped from almost twenty million sold in 2011 to some twelve million sold in 2014.[80] This trend away from dedicated e-readers is now supported by polling data as well as sales figures. In October 2015, Pew reported that only 19 percent of Americans owned a dedicated e-book reader, a sharp decline from the 32 percent reported in early 2014.[81] The 2016 Pew survey noted that only 8 percent of respondents had read a book using a dedicated e-reader. By contrast, 15 percent had read a book using a tablet and 13 percent had read a book on a cell phone.[82]

As the above passages indicate, there are growing indications that digital reading is increasingly being done on multipurpose devices such as tablets and even smartphones. In June 2014, for example, the website Mashable predicted that e-readers were destined to become the "next iPod," becoming redundant for most users due to tablets.[83] In August 2015, the *Wall Street Journal*, citing Nielsen data, reported that "the percentage of e-book buyers who read primarily on tablets was 41% in the first quarter of 2015, compared with 30% in 2012."[84] Yet the *Journal* argued that, in fact, "it's not the e-reader that will be driving future book sales, but the phone."[85] According to Nielsen, 54 percent of e-book buyers in 2014 used a smartphone for at least some of their reading, compared to 24 percent in 2012. In the words of one publishing executive, "The future of digital reading is on the phone."[86] Or, as Pew Research's Lee Rainie told *Publishers Weekly* in October 2016, "There is a big uptick in people using tablets and phones, and not so much dedicated e-book readers."[87]

Market research firm IHS noted back in 2012 where we were headed: "Single-task devices like the e-book are being replaced without remorse in the lives of consumers by their multifunction equivalents, in this case by media tablets."[88]

The leveling off of e-book and e-reader sales indicates that readers are not quite ready to completely forsake print. Surveys of reader preferences lend further credence to this interpretation. According to the latest annual Pew research study on book reading habits, published in September 2016, the percentage of those who read an e-book in the previous year has remained flat since 2014, the figure for both years being 28 percent, with a slight drop to 27 percent in 2015. To be certain, the percentage who indicated that they had read a print book has fluctuated a bit more, going from 69 percent in 2014 to 63 percent in 2015, then rebounding to 65 percent last year.[89] This relative stability likewise suggests that readers are not abandoning print for e-books. As Pew's Rainie told *Publishers Weekly*, "One of the things we hear when we talk to consumers about print books is that print is a fabulous technology. Ink on a page is amazingly portable, long-lasting, sharable. Print is still amazingly attractive to people. And, my general sense is that readers are happy with their pathways to books."[90]

Pew's findings are backed up by numerous other surveys of reader preferences. It seems to be a common belief that reading habits will change once "digital natives," those who have grown up with digital reading technology, come to the fore. Most of the evidence we have so far, however, does not indicate that this is happening. Contrary to conventional wisdom about "digital natives," it appears that even many millennials prefer print when engaged in intensive linear reading. In December 2013, Ricoh Americas Corporation reported that "most consumers do not see themselves giving up printed books, due to the benefits the physical form offers."[91] Among the study's findings were that nearly 70 percent of readers were unlikely to abandon print by 2016 and that "college students prefer printed textbooks to eBooks as they help students to concentrate on the subject matter at hand; electronic display devices such as tablet PCs tempt students to distraction."[92]

These findings are amply supported by sources both popular and scholarly. A February 2015 *Washington Post* article noted that "textbook makers, bookstore owners and college student surveys all say millennials still strongly prefer print for pleasure and learning, a bias that surprises reading experts given the same group's proclivity to consume most other content digitally."[93]

A 2010 user survey conducted by the University of California (UC) Libraries found that 44 percent of all respondents to the UC survey who had used e-books still preferred print, while 35 percent preferred digital texts.[94] This preference held for both students and faculty. Despite the shift from print to e-reading in the last decade, a number of studies of university students have found that the majority still prefer print books to digital, at least for certain purposes. For example, 53 percent of undergraduate respondents to the survey said that they preferred print books to electronic (27 percent preferred e-books).[95] As the survey report put it, "Many undergraduate respondents commented on the difficulty they have learning, retaining, and concentrating while in front of a computer."[96] A 2012 survey conducted at a college in Pennsylvania found that half the students twenty-two or younger preferred print to e-books. Among their reasons for preferring print were that it was "easier to focus on content/task at hand," "easier to absorb/comprehend information on paper rather than from a monitor," and "easier to remember content" in print than in digital format.[97] That same year, Staiger analyzed more than two dozen studies of e-book use in academic libraries. He found "a salient preference across all of the studies for physical books for extended or immersive reading."[98] More recently, Baron found in her survey conducted for *Words Onscreen* that the length of a text plays a major role in student reading preferences. Ninety-two percent of the American undergraduates she asked preferred print for long schoolwork texts, and 85 percent preferred print for long pleasure readings. The results for German and Japanese students were comparable. Baron summarized her findings as follows: "Does length matter when it comes to choice of reading platform? Absolutely. If the text is short, medium preference is not particularly strong—a mixture of hardcopy, digital screen, or no preference. Reading longer texts is an entirely different story."[99]

Returning to Pew's 2016 findings, they report that while 28 percent of U.S. adults read an e-book in the past year, just 6 percent read only e-books. By contrast, 38 percent read books exclusively in print, while 28 percent read in both formats. Interestingly, they found that only 6 percent of respondents in the eighteen-to-twenty-nine age group were e-book only readers. In Pew's words, "Young adults are no more likely than older adults to be 'digital-only' book readers."[100]

Perhaps the most fascinating recent study of reader preferences was an April 2016 survey of 4,992 book buyers by the Codex Group, a publishing industry research firm. In the words of *Publishers Weekly*, the Codex Group found that "e-book units purchased as a share of total books purchased fell from 35.9% in April 2015 to 32.4% in April 2016." This included "e-books published by traditional publishers and self-publishers and sold across all channels and in all categories."[101] Among other implications, this would suggest that independent e-book sales might not be making up for the decline in major publisher e-book sales, contrary to what some have argued. For our purposes, though, it is the Codex Group's findings on the demographics of print versus digital reading that are most interesting. According to the survey, 25 percent of book buyers expressed a desire to spend less time using electronic devices. It was the youngest demographic, eighteen-to twenty-four-year-olds, who most wanted to reduce their screen time, with 37 percent indicating such a desire. Nineteen percent of eighteen- to twenty-four-year-olds reported reading fewer e-books than before, again the highest response among all age groups. Of the total number of respondents reading fewer e-books (14 percent), 59 percent said it was because they preferred print. Perhaps the most telling piece of information from the survey was that, as *Publishers Weekly* put it, the "share of print books purchased was also the highest among the heaviest screen users, the so-called digital natives, ages 18–24 (83%), and lowest (61%) among 55-to-64-year-olds." In the words of Codex Group president Peter Hildick-Smith, a sort of "digital fatigue" seems to have set in among many millennials, manifested, among other ways, by a desire to pursue long-form linear reading primarily in print format.[102]

Even among K–12 students, whom one would think would be most receptive to primarily reading e-books, there is evidence that many still prefer print for certain forms of reading. According to Scholastic's 2015 Kids and Family Reading Report, 65 percent of kids between six and seventeen said that they would always want to read in print, an increase from 2012's 60 percent. Contrary to what you might expect, it is the youngest readers who are most likely to read in print; 84 percent of six- to eight-year-olds did most of their pleasure reading in print, compared to 62 percent of fifteen- to seventeen-year-olds [103]

In short, it appears likely that the current plateauing of e-book sales is not simply a result of major publishers manipulating prices. Nor can the growing body of reader survey data showing a continued desire for print texts, especially among younger readers, be easily dismissed. It does look to be more than a temporary blip reflecting cultural prejudices that will disappear as digital reading takes a deeper hold in society. Rather, digital skeptics would argue that all of this evidence taken together reflects an instinctive understanding of the differences between print reading and e-reading. Staiger, for example, believes that the research he describes "indicates that print books are preferred for what we typically think of as the kind of reading on which sustained intellectual inquiry depends, let alone the life of the mind."[104] Or, as one publishing executive told the *New York Times*, "People talked about the demise of physical books as if it was only a matter of time, but even 50 to 100 years from now, print will be a big chunk of our business."[105]

TILTING THE
BALANCE TO PRINT

I mentioned my own sympathy with the concerns of the digital skeptics at the beginning of this briefing. In my view, it is this evidence of what readers actually want, manifested in both sales data and expressed reader preferences, that tilts the balance in favor of those concerned about the fate of linear reading in the digital age. When you have a substantial percentage of tech-savvy millennial undergraduates, those you would least expect to be susceptible to what Thompson calls the "fabulous PR" accorded the print codex, making arguments for print that could be taken almost verbatim from Carr or Wolf, it is a powerful indicator that perhaps print and digital reading really are substantially different. The preference documented by Baron, albeit from a small sample of respondents, for reading long-form texts in print, seems to directly highlight the weakness of those studies supportive of digital reading's ability to foster similar comprehension to print, based as they are on readings of very short texts in controlled environments.

It is, of course, possible that Thompson, Jones, and other champions of digital reading are correct, and the current stabilized status of print reading is merely a brief blip in a virtually inevitable transition to a digital reading future. The expressed preferences of many readers for print when engaged in linear reading could simply be a vestigial cultural holdover, doomed to disappear as a new generation who has known digital devices their entire lives comes of age and show that there is no fundamental difference in reading format. The flatlining of e-book sales might just be a temporary correction, awaiting only this generational transition and dissipation of old prejudices to take off once again.

I'm not so sure, however. As awareness of the distractions inherent to many of our digital devices has come to permeate society, even some of those most critical of the digital skeptics have recently modified their views. While Thompson may have read *War and Peace* on his smartphone to prove a point, Bilton returned to reading books in print in 2013, citing as his reasons the lack of distractions as well as the tactile qualities of reading a physical book, which the work of Mangen has shown to help with comprehension.[106] Similarly, Shirky now refuses to allow his students to use digital technology in the classroom without permission due to its distracting effects. In his words, "The industry has committed itself to an arms race for my students' attention, and if it's me against Facebook and Apple, I lose."[107]

There are, of course, many proponents of the dedicated e-reader, such as Jacobs, who would object by saying that devices like the Kindle or Nook have the ability to bridge the reading divide. However, even if we assume that dedicated e-readers succeed in preserving the experience of deep reading in a digital container, there is still the question of how popular such devices will prove to be over the long-term. As I've shown above, the data strongly suggest that the dedicated e-reader is losing out to the multipurpose tablet and even the smartphone.

EVEN SOME OF THOSE MOST CRITICAL OF THE DIGITAL SKEPTICS HAVE RECENTLY MODIFIED THEIR VIEWS.

If tablets and/or smartphones do become the primary device for digital reading, with all their attendant possibilities of distraction and multitasking, it does not bode well for those who hope that deep reading can be preserved in the digital environment. The *New York Times* summarized the danger in March 2012: "People who read e-books on tablets like the iPad are realizing that while a book in print or on a black-and-white Kindle is straightforward and immersive, a tablet offers a menu of distractions that can fragment the reading experience, or stop it in its tracks."[108] Even if it is possible to engage in deep reading on a tablet, how many readers will choose to do so when Facebook or YouTube are just a click away, especially if their neural pathways have rewired themselves to want to seek the latter at the expense of the former?

There are, of course, those who would argue that the threat of distraction is nothing new. Postman, after all, warned of the threat to attention spans posed by television. Are not the present worries over reading in digital format simply a reflection of these earlier arguments? Unfortunately, in my view, this is an apples-and-oranges comparison. One can certainly make a case that the threat to our ability to engage in deep linear reading began with screen-based technologies such as movies and television. The main point to keep in mind, however, is that these technologies were completely self-contained and separate from the print codex. In the digital environment, by contrast, text, audio, and video are all brought to you on the same device. As Carr once famously observed, the Internet absorbs all previous information technologies and remakes them in its own image. In the digital environment, text ceases to be one distinct, self-contained format among many and becomes simply one form of content thoroughly integrated into the digital cornucopia. Television and film offered more stimulating alternatives while leaving the codex itself alone and unchanged. The digital information environment has granted the book no such luxury. That is why this time is indeed different. The threat to deep linear reading may not have originated in the digital age, but it has certainly greatly accelerated because of it.

This especially applies to the corollary argument that is often heard: How is the current threat of digital distraction any greater when reading an e-book versus a print codex? Isn't it just as easy to put down a print book and pick up a tablet or smartphone as it is to close out your e-reading app and start browsing Facebook? The answer, in my view, is no, and again I return to neuroplasticity. The digital environment is literally rewiring our brains to seek stimulative, short-term gratification at the expense of our ability to think and read in depth. In this situation, how much more challenging is it to read at length on the very same screen from which your brain expects quick scanning, 140-character tweets, and amusing cat videos than it is to read from a printed page or on a dedicated e-reader that does not offer such opportunities for distraction?

Thus *the digital reading environment offers not a difference in degree but a difference in kind, one that is transformational in nature rather than evolutionary.* As the digital age unfolds, it is likely to *substantially alter* both the nature of reading and the nature of the book itself as deep linear reading fades in importance and functional tabular reading becomes more widespread than ever. This will in turn alter the way people write and even the ways they think, leading to a likely decline of deep analytical thought for the purpose of forming broad conceptual frameworks in favor of a more immediate, purely functional form of decision-oriented thinking based on rapidly acquired snippets of information.

CONCLUSION
Building the "Bi-literate Brain"

Digital reading is here to stay. No one, obviously, is calling to turn back the clock. One need only browse the URL-laden endnotes of this briefing to see how even a digital skeptic like me has become dependent on screen-based reading. The key question, as noted at the beginning of this briefing, then becomes, How do we, as librarians, publishers, and software creators, work to preserve reading in all its richness in the digital age?

To start with, we must avoid the intellectual trap that technology blogger Michael Sacasas has termed the "Borg complex": the belief that newer technologies are, by definition, inherently superior to preexisting ones and that all technological change is inevitable, so arguing about it is pointless.[109] Yale computer scientist David Gelernter made a similar point in an interview with NPR: "It's not as if books have lost an argument. The problem is there hasn't been an argument. Technology always gets a free pass. [People] take it for granted that if the technology is new it must be better."[110]

Obviously, many will make the fair objection that technology, especially in the case of the e-book, has not received a "free pass." If there wasn't a robust debate over technology and the nature of reading, this briefing wouldn't exist. Yet when you step back and take a broader view of the history of technology, Gelernter has a point. For all their eloquence and the favorable reception their arguments receive in certain circles, it is obvious that technology critics and analysts such as Mumford, McLuhan, and Postman have had a negligible impact. To the extent that digital reading skeptics such as Carr and

Baron have influenced the current reversion to print as a preferred format for long-form linear reading, it is because their arguments have expressed what many contemporary readers have themselves experienced in the digital environment.

This continued preference for print for the linear mode of reading is so widespread and deeply felt that it seems unfair to dismiss it merely as a vestigial cultural construct. Even if it is cultural in origin, surely the very fact of its depth and breadth might indicate that there are good reasons for its prevalence. As Gelernter notes, part of the reason the print codex retains so much of its popularity is its simple, elegant, user-friendly nature: "'It's an inspiration of the very first order. It's made to fit human hands and human eyes and human laps in the way that computers are not,' he says, wondering aloud why some are in such a rush to discard a technology that has endured for centuries."[111] Even assuming that digital reading devices can one day truly mimic all the features that made the print codex such a beloved, enduring technology, would doing so not become, in part, simply an exercise in reinventing the wheel?

This widely held desire to preserve the ability for deep linear reading has manifested itself in several ways. One notable development is the rise of what's been called the "Slow Reading Movement." A September 2014 *Wall Street Journal* profile offered this useful definition of slow reading: "Slow reading means a return to a continuous, linear pattern, in a quiet environment free of distractions. Advocates recommend setting aside at least 30 to 45 minutes in a comfortable chair far from cell-phones and computers. Some suggest scheduling time like an exercise session. Many recommend taking occasional notes to deepen engagement with the text."[112]

The idea of slow reading has been around for several years. Maura Kelly, writing for *The Atlantic*, issued this March 2012 call for a "Slow Books Movement" along remarkably similar lines:

Aim for 30 minutes a day. You can squeeze in that half hour pretty easily if only, during your free moments—whenever you find yourself automatically switching on that boob tube, or firing up your laptop to check your favorite site, or scanning Twitter for something to pass the time—you pick up a meaningful work of literature. Reach for your e-reader, if you like. The Slow Books movement won't stand opposed to technology on purely nostalgic or aesthetic grounds. (Kindles et al. make books like *War and Peace* less heavy, not less substantive, and also ensure you'll never lose your place.)[113]

Noteworthy in Kelly's description is that she considers dedicated e-readers to be suitable tools for deep linear reading as well as print. This point is also noted in the *Wall Street Journal* piece: "Some hard-core proponents say printed books are best, in part because they're more visible around the house and serve as a reminder to read. But most slow readers say e-readers and tablets are just fine, particularly if they're disconnected from the Internet."[114] As Pew's Rainie puts it, "Our data are very clear that there is a class of Americans who just can't get enough books, and if they can't be with the format they love, they love the format they're with."[115]

This observation is important in several ways. For one thing, should the Slow Reading Movement become widespread enough, it could serve as a lifeline for the dedicated e-reading device, allowing it to retain some degree of market share and cultural traction. More profoundly, perhaps, it suggests that digital devices are suitable for deep linear reading when properly designed and when the user is enabled, and willing, to avoid distractions. The problem, as we have seen, is that most digital devices are seemingly engineered to foster distraction, to seize "our attention only to scatter it," in Carr's words.[116]

Another factor to keep in mind regarding the Slow Reading Movement is that it seems very much a middle-class bourgeois bohemian phenomenon. As such, it is a product of what Northwestern University sociologist Wendy Griswold has described as "a self-perpetuating minority that I have called the reading class."[117] The emerging outlines of such a class are already visible in the data from Pew and others. It disproportionately comprises such elements as college graduates, young adults, and women.

On the one hand, this makes it difficult to determine how much of a societal impact this movement in support of deep linear reading will have in the face of the tremendous growth of the digital information environment. For example, literary reading, according to the National Endowment for the Arts, has fallen to the lowest level ever recorded, with just 43 percent of adults in 2015 having read at least one piece of literature in the last year. The figure was

57 percent in 1982, when this question was first surveyed. Sixty-eight percent of those with a graduate education were literary readers in 2015 versus only 30 percent of those with a high school diploma.[118]

On the other hand, the "reading" class has enabled book reading to remain relatively stable in the last few years in the face of an ever-proliferating variety of digital entertainment options. In Rainie's words, "With so many ways people can allocate their time now, I think the surprising thing for us is that books are holding their own."[119]

Slow readers are disproportionate users of libraries, heavy purchasers of books in both print and electronic formats, and the key hope for maintaining some notable form of dedicated e-reader market. As such, librarians, publishers, and e-reader designers would do well to be aware of this movement and take the preferences of its members into account.

At heart, the Slow Reading Movement is a spontaneous, grassroots effort to preserve what Wolf has called the "bi-literate brain," one equally conversant in both digital tabular reading and long-form linear reading. Wolf briefly explained to the *Washington Post* what this would entail: "We can't turn back. . . . We should be simultaneously reading to children from books, giving them print, helping them learn this slower mode, and at the same time steadily increasing their immersion into the technological, digital age. It's both. We have to ask the question: What do we want to preserve?"[120]

In an interview with *The New Yorker*, Wolf expressed her confidence that, in *The New Yorker*'s words, "we can learn to navigate online reading just as deeply as we once did print—if we go about it with the necessary thoughtfulness."[121] The piece goes on to describe her efforts to implement this vision of a biliterate brain: "The same plasticity that allows us to form a reading circuit to begin with, and short-circuit the development of deep reading if we allow it, also allows us to learn how to duplicate deep reading in a new environment. . . . We cannot go backwards. As children move more toward an immersion in digital media, we have to figure out ways to read deeply there."[122]

Ultimately, this vision of a biliterate future, combining print and digital in a way that enables and integrates the best features of both, enabling both linear and tabular reading, is what all of us involved in reading need to work toward. In practice, for the foreseeable future, this means ensuring a continued place for the print codex in the digital age.

This recommendation may come as a bit of a surprise in light of the preceding passages. After all, as we've seen, the Slow Reading Movement is very much in favor of the Kindle and similar e-reading devices, while a world-renowned expert on reading and neuroplasticity such as Wolf firmly believes that we can discover how to make linear reading viable in the digital realm. These factors suggest that print as a format is not necessarily indispensable to long-form linear reading.

It is true, as we have also seen, that the print/linear versus digital/tabular framework is far from precise. Obviously, a great deal of tabular reading has been, and continues to be, done in print format, and it is not impossible to engage in linear reading on a digital device, especially a dedicated e-reader. Another key factor involves differences among types of literature. Genre fiction, for example, seems much better suited to e-reading than do monographs in history or philosophy. There are also differences among academic disciplines, with the humanities placing far greater emphasis on linear reading of lengthy texts than do the STEM fields. Finally, it is important to keep in mind individual preferences. The current reading environment is not a one-size-fits-all situation.

Having taken these factors into account, the distinction I would make is that the print codex fosters—indeed, is expressly designed to facilitate—the ability to read in depth and at length in a way that most current digital devices do not. No one needs to modify the paper book to make it suitable for long-form linear reading. The print codex has shaped the way we read, the way we write, and the way we think for centuries. Our society continues to live off of the accumulated cultural capital of print literacy. If we marginalize print, we risk marginalizing an entire way of reading, writing, and thinking that has proved heretofore indispensable to our society, with potentially serious consequences. Just as the advent of the radio did not do away with the record player and television did not end the movie theater, so there is no reason why screen-based reading should spell the end of print reading. Just as the record player and the movie theater continued to fill very specific needs and functions that the radio and television could not, so the print codex serves as an ideal mechanism for in-depth, distraction-free linear reading in a way that the most popular digital devices do not. With a substantial body of scholarly and popular opinion now seemingly in agreement on the need for long-form linear reading and the dangers of digital distraction, discarding a proven centuries-old technology ideal for meeting those qualifications seems extremely foolhardy.

Instead of being seen as interchangeable, print and digital should be seen as complementary formats for text, both of which are necessary. We need to move beyond the simple dichotomy of print versus digital and understand that both formats are indispensable going forward. Instead of print or digital, let us think of print *and* digital.

NOTES

1 Barry W. Cull, "Reading Revolutions: Online Digital Text and Implications for Reading in Academe," *First Monday* 16, no. 6 (2011), accessed July 29, 2013, http://firstmonday.org/ojs/index.php/fm/article/view/3340/2985.

2 See David M. Durant and Tony Horava, "The Future of Reading and Academic Libraries," *portal: Libraries and the Academy* 15, no. 1 (2015): 5–27; and David M. Durant, "Resistance Is Not Futile: Why Print Collections Still Matter in the Digital Age," *Against the Grain* 27, no. 3 (2015): 26–28.

3 Maryanne Wolf and Mirit Barzillai, "The Importance of Deep Reading," *Educational Leadership* 66, no. 6 (2009), accessed July 29, 2013, http://ase.tufts.edu/crlr/documents/2009EL-ImportanceDeepReading.pdf.

4 Christopher Rowe, "The New Library of Babel?," *First Monday* 18, no. 2 (2013), accessed July 29, 2013, http://firstmonday.org/htbin/cgiwrap/bin/ojs/index.php/fm/article/view/3237/3416.

5 N. Katherine Hayles, "How We Read: Close, Hyper, Machine," *ADE Bulletin* 150 (2010), accessed August 24, 2015, http://nkhayles.com/how_we_read.html.

6 Ibid.

7 Wolf and Barzillai, "The Importance of Deep Reading."

8 Ibid.

9 For a good overview, see Kevin Kelly, "Reading in a Whole New Way," *Smithsonian Magazine*, July–August 2010, accessed July 30, 2013, http://www.smithsonianmag.com/specialsections/40th-anniversary/Reading-in-a-Whole-New-Way.html.

10 Sven Birkerts, *The Gutenberg Elegies: The Fate of Reading in an Electronic Age* (New York: Faber & Faber, 2006), 154.

11 Nicholas Carr, "Is Google Making Us Stupid?," *The Atlantic*, July–August 2008, accessed August 27, 2015, http://www.theatlantic.com/magazine/archive/2008/07/is-google-making-us-stupid/306868/.

12 Ibid.

13 Ibid.

14 Ibid.

15 Nicholas Carr, *The Shallows: What the Internet Is Doing to Our Brains* (New York: Norton, 2010).

16 Ibid., 118.

17 Gary W. Small, Teena D. Moody, Prabha Siddarth, and Susan Y. Bookheimer, "Your Brain on Google: Patterns of Cerebral Activation during Internet Searching," *American Journal of Geriatric Psychiatry* 17, no. 2 (2009): 116–26, accessed July 29, 2013, doi:10.1097/JGP.0b013e3181953a02.

18 See Carr, *The Shallows*, 120–26.

19 Jakob Nielsen, "F-Shaped Pattern for Reading Web Content," *Jakob Nielsen's Alertbox*, April 17, 2006, accessed July 29, 2013, http://www.nngroup.com/articles/f-shaped -pattern-reading-web-content/.

20 *Information Behaviour of the Researcher of the Future* (London: University College, 2008): 10, accessed July 28, 2013, http://www.educause.edu/library/resources/ information-behaviour-researcher-future.

21 Wolf and Barzillai, "The Importance of Deep Reading."

22 Hayles, "How We Read."

23 Ferris Jabr, "The Reading Brain in the Digital Age: The Science of Paper versus Screens," *Scientific American*, April 11, 2013, accessed August 28, 2015, http://www .scientificamerican.com/article/reading-paper-screens/.

24 Naomi S. Baron, *Words Onscreen: The Fate of Reading in a Digital World* (Oxford: Oxford University Press, 2015), 39.

25 Ibid., 88.

26 Michael S. Rosenwald, "Serious Reading Takes a Hit from Online Scanning and Skimming, Researchers Say," *Washington Post*, April 6, 2014, accessed August 29, 2015, http://www .washingtonpost.com/local/serious-reading-takes-a-hit-from-online-scanning-and-skimming -researchers-say/2014/04/06/088028d2-b5d2-11e3-b899-20667de76985_story.html.

27 Ibid.

28 Susan Greenfield, *Mind Change: How Digital Technologies Are Leaving Their Mark on Our Brains* (New York: Random House, 2015), 233.

29 Maryanne Wolf, "Our 'Deep Reading' Brain: Its Digital Evolution Poses Questions," *Nieman Reports*, Summer 2010, accessed August 28, 2015, http://niemanreports.org/ articles/our-deep-reading-brain-its-digital-evolution-poses-questions/.

30 Baron, *Words Onscreen*, 39.

31 Ibid.

32 Diane Wachtell, "Books Aren't Crucial, but Long-Form Texts Are," *The Chronicle Review*, September 26, 2010, accessed August 29, 2015, http://chronicle.com/article/ Books-Arent-Crucial-but/124569/.

33 Ibid.

34 Nick Bilton, *I Live in the Future & Here's How It Works* (New York: Crown Business, 2010), 48.

35 Ibid., 136.

36 Ibid.

37 Clive Thompson, "Reading *War and Peace* on My iPhone," *Book Riot*, 2015, accessed August 29, 2015, http://bookriot.com/quarterly/bkr07/.

38 John Jones, "Book Lacks Digital Reading Details," *DML Central*, July 20, 2015, accessed August 29, 2015, http://dmlcentral.net/book-lacks-digital-reading-details/.

39 Alan Jacobs, *The Pleasures of Reading in an Age of Distraction* (Oxford: Oxford University Press, 2011), 82.

40 Ibid.

41 George Stachokas. *After the Book: Information Services for the 21st Century* (London: Chandos Publishing, 2014), 16.

42 Mathew Ingram, "No, E-book Sales Are Not Falling, despite What Publishers Say," *Fortune*, September 24, 2015, accessed December 16, 2015, http://fortune.com/2015/09/24/ebook-sales/.

43 Quoted in Rachel Nuwer, "Are Paper Books Really Disappearing?," *BBC*, January 25, 2016, accessed February 9, 2017, http://www.bbc.com/future/story/20160124-are-paper-books-really-disappearing.

44 Ibid.

45 Marc Prensky, "In the 21st-Century University, Let's Ban (Paper) Books," *The Chronicle of Higher Education*, November 13, 2011, accessed December 16, 2015, http://chronicle.com/article/In-the-21st-Century/129744/.

46 Christine Rosen, "People of the Screen," *The New Atlantis*, Fall 2008, accessed August 7, 2017, http://www.thenewatlantis.com/publications/people-of-the-screen.

47 For example, see Clay Shirky, *Cognitive Surplus: Creativity and Generosity in a Connected Age* (New York: Penguin, 2010).

48 Quoted in Nicholas Carr, "Containers and Their Contents," *Rough Type* (blog), January 3, 2013, accessed July 30, 2013, http://www.roughtype.com/?p=2315.

49 Kelly, "Reading in a Whole New Way."

50 Quoted in Nuwer, "Are Paper Books Really Disappearing?"

51 Kevin Kelly, "Scan This Book!," *New York Times Magazine*, May 14, 2006, accessed July 31, 2013, http://www.nytimes.com/2006/05/14/magazine/14publishing.html.

52 Nicholas Carr, "The Remains of the Book," *Rough Type* (blog), September 30, 2011, accessed July 31, 2013, http://www.roughtype.com/archives/2011/09/the_seethrough_1.php.

53 Jeff Staiger, "How E-books Are Used: A Literature Review of the E-book Studies Conducted from 2006 to 2011," *Reference & User Services Quarterly* 51, no. 4 (2012): 361, accessed July 29, 2013, doi:10.5860/rusq.51n4.355.

54 Jakob Nielsen, "iPad and Kindle Reading Speeds," *Jakob Nielsen's Alertbox*, July 2, 2010, accessed July 31, 2013, http://www.nngroup.com/articles/ipad-and-kindle-reading-speeds/.

55 F. Kretzschmar, D. Pleimling, J. Hosemann, S. Füssel, I. Bornkessel-Schlesewsky, and M. Schlesewsky, "Subjective Impressions Do Not Mirror Online Reading Effort: Concurrent EEG-Eyetracking Evidence from the Reading of Books and Digital Media," *PLoS ONE* 8, no. 2 (2013): e56178, accessed August 30, 2015, doi:10.1371/journal.pone.0056178.

56 Ibid.

57 Ibid.

58 Sara J. Margolin, Casey Driscoll, Michael J. Toland, and Jennifer Little Kegler, "E-readers, Computer Screens, or Paper: Does Reading Comprehension Change across Media Platforms?," *Applied Cognitive Psychology* 27 (2013): 512–19.

59 Anne Mangen, Bente R. Walgermo, and Kolbjørn Brønnick, "Reading Linear Texts on Paper versus Computer Screen: Effects on Reading Comprehension," *International Journal of Educational Research* 58 (2013): 61–68, doi:10.1016/j.ijer.2012.12.002.

60 Alison Flood, "Readers Absorb Less on Kindles than on Paper, Study Finds," *The Guardian*, August 19, 2014, accessed August 31, 2015, http://www.theguardian.com/books/2014/aug/19/readers-absorb-less-kindles-paper-study-plot-ereader-digitisation.

61 Ibid.

62 Hayles, "How We Read."

63 Jeremy Greenfield, "Ebooks Will Make Us Dumber, or They Won't," *Forbes*, August 20, 2014, accessed August 31, 2015, http://www.forbes.com/sites/jeremygreenfield/2014/08/20/will-ebooks-make-us-dumber/.

64 Kathryn Zickuhr and Lee Rainie, "E-reading Rises as Device Ownership Jumps," *Pew Internet & American Life Project*, January 16, 2014, accessed August 31, 2015, http://www.pewinternet.org/2014/01/16/e-reading-rises-as-device-ownership-jumps/.

65 Ibid.

66 Claire Cain Miller and Julie Bosman, "E-books Outsell Print Books at Amazon," *New York Times*, May 19, 2011, accessed July 28, 2013, http://www.nytimes.com/2011/05/20/technology/20amazon.html.

67 "E-book Sales Are up 43%, but That's Still a 'Slowdown,'" *USA Today*, May 16, 2013, accessed July 28, 2013, http://www.usatoday.com/story/life/books/2013/05/15/e-book-sales/2159117/.

68 Rick Anderson, "Print on the Margins: Circulation Trends in Major Research Libraries," *Library Journal*, June 2, 2011, accessed March 31, 2015, http://lj.libraryjournal.com/2011/06/academic-libraries/print-on-the-margins-circulation-trends-in-major-research-libraries/.

69 Alexandra Alter, "The Plot Twist: E-book Sales Slip, and Print Is Far from Dead," *New York Times*, September 22, 2015, accessed December 16, 2015, http://www.nytimes.com/2015/09/23/business/media/the-plot-twist-e-book-sales-slip-and-print-is-far-from-dead.html.

70 Ibid.

71 Jeremy Greenfield, "Study: Ebook Growth Stagnating in 2013," *Digital Book World*, October 30, 2013, accessed March 31, 2015, http://www.digitalbookworld.com/2013/study-ebook-growth-stagnating-in-2013/. Cited by Nicholas Carr, "Peak Ebook?," *Rough Type* (blog), November 7, 2013, accessed March 31, 2015, http://www.roughtype.com/?p=3966.

72 "Children's and Young Adult Hardback Books up 95.2% in July 2016 vs. July 2015," *Association of American Publishers*, December 20, 2016, accessed February 13, 2017, http://newsroom.publishers.org/childrens-and-young-adult-hardback-books-up-952-in-july-2016-vs-july-2015/.

73 Ingram, "E-book Sales."

74 Kate Stoltzfus, "Do 'Digital Natives' Prefer Paper Books to E-books?," *Education Week*, November 8, 2016, accessed February 13, 2017, http://www.edweek.org/ew/articles/2016/11/09/do-digital-natives-prefer-paper-books-to.html.

75 Andrew Richard Albanese, "Print or Digital, It's Reading That Matters," *Publishers Weekly*, September 16, 2016, accessed February 12, 2017, http://www.publishersweekly.com/pw/by-topic/digital/content-and-e-books/article/71500-print-digital-and-what-really-matters.html.

76 Alter, "The Plot Twist."

77 Padraig Belton and Matthew Wall, "Did Technology Kill the Book or Give It New Life?," *BBC*, August 14, 2015, accessed August 31, 2015, http://www.bbc.com/news/business-33717596.

78 Quoted in Alison Flood, "Ebook Sales Falling for the First Time, Finds New Report," *The Guardian*, February 3, 2016, accessed February 12, 2017, https://www.theguardian.com/books/2016/feb/03/ebook-sales-falling-for-the-first-time-finds-new-report.

79 Ibid.

80 Alter, "The Plot Twist."

81 Monica Anderson, "Technology Device Ownership: 2015," *Pew Research Center*, October 29, 2015, accessed December 18, 2015, http://www.pewinternet.org/2015/10/29/technology-device-ownership-2015/.

82 Andrew Perrin, "Book Reading 2016," *Pew Research Center*, September 1, 2016, accessed February 11, 2017, http://www.pewinternet.org/2016/09/01/book-reading-2016/.

83 Todd Wasserman, "Why E-readers Are the Next iPods," *Mashable*, June 27, 2014, accessed August 31, 2015, http://mashable.com/2014/06/27/e-readers-next-ipods/?utm_cid=mash-com-Tw-main-link.

84 Jennifer Maloney, "The Rise of Phone Reading," *Wall Street Journal*, August 14, 2015, accessed February 12, 2017, https://www.wsj.com/articles/the-rise-of-phone-reading-1439398395.

85 Ibid.

86 Ibid.

87 Andrew Albanese, "Frankfurt Book Fair 2016: PW Talks to Pew Research Center's Lee Rainie about Reading in the Digital Age," *Publishers Weekly*, October 21, 2016, accessed February 12, 2017, http://www.publishersweekly.com/pw/by-topic/international/Frankfurt-Book-Fair/article/71787-frankfurt-book-fair-2016-pw-talks-to-pew-research-center-s-lee-rainie-about-reading-in-the-ditgital-age.html.

88 Jordan Selburn, "Ebook Readers: Device to Go the Way of Dinosaurs?," *IHS iSuppli*, December 10, 2012, accessed December 18, 2015, https://technology.ihs.com/417568/ebook-readers-device-to-go-the-way-of-dinosaurs. Cited by Nicholas Carr, "E-reading after the E-reader," *Rough Type* (blog), December 30, 2012, accessed July 31, 2013, http://www.roughtype.com/?p=2245.

89 Perrin, "Book Reading 2016."

90 Quoted in Albanese, "Frankfurt Book Fair 2016."

91 "New Research Reveals Unexpected Positive Outlook for the Printed Book, Due to Love of the Medium," *Ricoh Americas Corporation*, December 9, 2013, accessed December 18, 2015, https://www.ricoh-usa.com/news/news_release.aspx?prid=1164&alnv=pr.

92 Ibid.

93 Michael S. Rosenwald, "Why Digital Natives Prefer Reading in Print. Yes, You Read That Right," *Washington Post*, February 22, 2015, accessed August 31, 2015, http://www.washingtonpost.com/local/why-digital-natives-prefer-reading-in-print-yes-you-read-that-right/2015/02/22/8596ca86-b871-11e4-9423-f3d0a1ec335c_story.html.

94 Chan Li, Felicia Poe, Michele Potter, Brian Quigley, and Jacqueline Willis, "UC Libraries Academic E-book Usage Survey: Springer E-book Pilot Project," *University*

of California Libraries (May 2011), 12, accessed July 31, 2013, http://www.cdlib.org/services/uxdesign/docs/2011/academic_ebook_usage_survey.pdf. Cited by Nicholas Carr, "Another Study Points to Advantages of Printed Textbooks," *Rough Type* (blog), June 27, 2011, accessed July 31, 2013, http://www.roughtype.com/?p=1496.

95 Ibid.

96 Li, Poe, Potter, Quigley, and Willis, "UC Libraries," 11.

97 Beth Jacoby, e-mail to COLLDV-L mailing list, November 17, 2012, accessed July 31, 2013, http://serials.infomotions.com/colldv-l/archive/2012/201211/0221.html.

98 Staiger, "How E-books Are Used," 362.

99 Baron, *Words Onscreen*, 85.

100 Perrin, "Book Reading 2016."

101 Jim Milliot, "As E-book Sales Decline, Digital Fatigue Grows," *Publishers Weekly*, June 17, 2016, accessed February 13, 2017, http://www.publishersweekly.com/pw/by-topic/digital/retailing/article/70696-as-e-book-sales-decline-digital-fatigue-grows.html.

102 Ibid.

103 Stoltzfus, "Do 'Digital Natives' Prefer Paper Books?"

104 Staiger, "How E-books Are Used," 362.

105 Quoted in Alter, "The Plot Twist."

106 Nick Bilton, "The Allure of the Print Book," *Bits* (blog), *New York Times,* December 2, 2013, accessed March 27, 2015, http://bits.blogs.nytimes.com/2013/12/02/the-print-book-here-to-stay-at-least-for-now/.

107 Clay Shirky, "Why I Just Asked My Students to Put Their Laptops Away," *Medium*, September 8, 2014, accessed August 31, 2015, https://medium.com/@cshirky/why-i-just-asked-my-students-to-put-their-laptops-away-7f5f7c50f368.

108 Julie Bosman and Matt Richtel, "Finding Your Book Interrupted . . . by the Tablet You Read It On," *New York Times*, March 4, 2012, accessed July 31, 2013, http://www.nytimes.com/2012/03/05/business/media/e-books-on-tablets-fight-digital-distractions.html?_r=1&hp.

109 Michael Sacasas, "Borg Complex: A Primer," *The Frailest Thing*, March 1, 2013, accessed March 1, 2015, http://thefrailestthing.com/2013/03/01/borg-complex-a-primer/.

110 Quoted in Eric Weiner, "Technology of Books Has Changed, but Bookstores Are Hanging in There," *NPR*, May 28, 2015, accessed August 31, 2015, http://www.npr.org/2015/05/28/408787099/the-technology-of-books-has-changed-but-bookstores-are-hanging-in.

111 Ibid.

112 Jeanne Whalen, "Read Slowly to Benefit Your Brain and Cut Stress," *Wall Street Journal*, September 16, 2014, accessed August 7, 2017, https://www.wsj.com/articles/read-slowly-to-benefit-your-brain-and-cut-stress-1410823086.

113 Maura Kelly, "A Slow-Books Manifesto," *The Atlantic*, March 26, 2012, accessed August 31, 2015, http://www.theatlantic.com/entertainment/archive/2012/03/a-slow-books-manifesto/254884/.

114 Whalen, "Read Slowly."

115 Quoted in Albanese, "Frankfurt Book Fair 2016."

116 Carr, *The Shallows*, 118.

117 Wendy Griswold, *Regionalism and the Reading Class* (Chicago: University of Chicago Press, 2008), 66.

118 Christopher Ingraham, "The Long, Steady Decline of Literary Reading," *Wonkblog* (blog), *Washington Post*, September 7, 2016, accessed February 13, 2017, https://www .washingtonpost.com/news/wonk/wp/2016/09/07/the-long-steady-decline-of-literary -reading/.

119 Quoted in Albanese, "Frankfurt Book Fair 2016."

120 Rosenwald, "Serious Reading Takes a Hit."

121 Maria Konnikova, "Being a Better Online Reader," *The New Yorker*, July 16, 2014, accessed August 31, 2015, http://www.newyorker.com/science/maria-konnikova/being -a-better-online-reader.

122 Ibid.

ABOUT THE AUTHOR

David M. Durant is Associate Professor and Federal Documents and Social Sciences Librarian at J. Y. Joyner Library, East Carolina University in Greenville, North Carolina. He holds a master's of science degree in library and information services from the School of Information, University of Michigan, and an MA in history from the University of California, Los Angeles. He has published articles in *portal, Library Journal, The Chronicle of Higher Education*, and *Against the Grain*. He has written book reviews for *Against the Grain* and for *Choice*.

ORCID ID: 0000-0002-3328-0235

www.ingramcontent.com/pod-product-compliance
Lightning Source LLC
Chambersburg PA
CBHW081251040426

42452CB00015B/2786